THE SMALLEST DINOSAURS

BY **"DINO" DON LESSEM**

ILLUSTRATIONS BY **JOHN BINDON**

LERNER PUBLICATIONS COMPANY / MINNEAPOLIS

Dedicated to Guan Jian, friend of dinosaurs, big and small

Text copyright © 2005 by Dino Don, Inc.
Illustrations copyright © 2005 by John Bindon
Reprinted in 2006
Photographs courtesy of Don Lessem, Dino Don, Inc., p. 12 (inset); Elaine Kennedy, pp. 12–13.

This book is available in two editions:
Library binding by Lerner Publications Company,
    a division of Lerner Publishing Group
Soft cover by First Avenue Editions,
    an imprint of Lerner Publishing Group
241 First Avenue North
Minneapolis, MN 55401 U.S.A.

Website address: www.lernerbooks.com

Library of Congress Cataloging-in-Publication Data

Lessem, Don.
    The smallest dinosaurs / by Don Lessem ; illustrations by John Bindon.
        p.   cm. — (Meet the dinosaurs)
    Includes index.
    Contents: Meet the smallest dinosaurs — Smaller than you'd think — Living small — Small, smaller, smallest — Where are they?
    ISBN-13: 978-0-8225-1372-8 (lib. bdg. : alk. paper)
    ISBN-10: 0-8225-1372-2 (lib. bdg. : alk. paper)
    ISBN-13: 978-0-8225-2575-2 (pbk. : alk. paper)
    ISBN-10: 0-8225-2575-5 (pbk. : alk. paper)
    Dinosaurs—Juvenile literature. (1. Dinosaurs.) I. Bindon, John, ill. II. Title.
QE861.5.L48 2005
567.9—dc22   2003017113

Printed in China
2 3 4 5 6 7 – DP – 11 10 09 08 07 06

TABLE OF CONTENTS

MEET THE SMALLEST DINOSAURS

WELCOME, DINOSAUR FANS!

I'm "Dino" Don. I *love* dinosaurs. We don't think of dinosaurs as being small. But most dinosaurs were smaller than a car. In fact, the smallest were as little as a crow! Want to meet some of the tiny dinosaurs? Here are some quick facts on them:

COMPSOGNATHUS **(KOMP-sohg-NAY-thuhs)**
Length: 4.5 feet
Home: western Europe
Time: 151 million years ago

EPIDENDROSAURUS **(EHP-ee-DEHN-droh-SAWR-uhs)**
Length: 2.8 feet
Home: Asia
Time: 124 million years ago

LESOTHOSAURUS **(luh-SOH-toh-SAWR-uhs)**
Length: 3 feet
Home: southern Africa
Time: 200 million years ago

LIAOCERATOPS **(LYOW-SAYR-uh-tahps)**
Length: 3 feet
Home: Asia
Time: 76 million to 72 million years ago

LIGABUEINO **(LEE-gah-BOO-ay-EE-noh)**
Length: 2.5 feet
Home: southeastern South America
Time: 127 million years ago

MICRORAPTOR **(MY-kroh-RAP-tohr)**
Length: 1.8 feet
Home: Asia
Time: 124 million years ago

PARVICURSOR **(PAHR-vih-KUHR-sohr)**
Length: 3.3 feet
Home: northern Asia
Time: 84 million years ago

SHUVUUIA (shu -VOO-ee-uh)
Length: 3.3 feet
Home: northern Asia
Time: 83.5 million years ago

SMALLER THAN YOU'D THINK

A mousy animal is on the run. It is running from a little hunter the size of a chicken. The killer dinosaur is *Parvicursor*. It is a fast runner. But it is not fast enough to catch the mousy animal.

Parvicursor turns and sees a tiny lizard.
It snaps up the lizard in its jaws. With its
sharp teeth, the dinosaur crushes the
lizard and swallows it whole.

THE TIME OF THE SMALLEST DINOSAURS

Lesothosaurus

Compsognathus

200 million
years ago

151 million
years ago

Many dinosaurs were huge. But not all
were big. Like all dinosaurs, tiny dinosaurs
had special skeletons. They had necks that
curved like the letter *S*. And their skeletons
let them stand up straight.

Microraptor

Parvicursor

Liaoceratops

124 million
years ago

84 million
years ago

76 million
years ago

So why did some dinosaurs grow big, but some did not? We don't know. Some were giants and chewed on big plants. Others nibbled at tiny bushes. Some meat eaters were huge. Others were tiny and chased even smaller animals and insects.

DINOSAUR FOSSIL FINDS

The numbers on the map on page 11 show some of the places where people have found fossils of the dinosaurs in this book. You can match each number on the map to the name and picture of the dinosaurs on this page.

1. Compsognathus   2. Epidendrosaurus   3. Lesothosaurus   4. Liaoceratops

5. Ligabueino   6. Microraptor   7. Parvicursor   8. Shuvuuia

We know about small dinosaurs from **fossils,** or the traces they left behind. Bones, footprints, and the marks left in rock by plants from long ago are examples of fossils.

Some small dinosaur fossils were first found in Germany at the Eichstatt quarry. People dig up stone for buildings at this quarry. The **limestone** there is 145 million years old. It is very smooth and keeps tiny skeletons perfectly.

More than 100 years ago, workers in the
Eichstatt quarry broke open a stone.
Inside was the complete skeleton of a
tiny dinosaur. It was only 3 feet long!

Scientists named the dinosaur *Compsognathus*.
It is one of the smallest dinosaurs known.

LIVING SMALL

Many tiny dinosaurs were meat eaters. But they could not fight with big dinosaurs for food. Tiny meat eaters often had to find food where other dinosaurs could not reach.

Epidendrosaurus is one of the strangest-looking dinosaurs. Its very long fingers and toes may have helped it to climb into trees to hunt.

Like some tiny meat eaters, *Shuvuuia* had huge eyes, a big brain, strong fingers, and quick feet. It also may have had feathers. The feathers weren't for flying. But they would have kept *Shuvuuia* warm on cold desert nights.

Shuvuuia also had a sharp beak that could bend. We aren't sure how *Shuvuuia* used its beak. But it probably helped this little hunter survive in a world of giants. For example, its beak might have been useful for prying insects or lizards out of holes.

Lesothosaurus had a long tail. But its body was only as big as a house cat. It lived among big dinosaurs and tough plants that made no fruits, flowers, or other food. How could *Lesothosaurus* find enough food and keep safe?

Lesothosaurus was a fast runner with good eyesight. It could see hunters and escape from them. Its long snout and sharp beak may have helped it to snip plants or maybe eat insects and lizards. Its thumb probably helped the little dinosaur grip its food.

Little *Ligabueino* is looking for smaller animals
to eat. Suddenly, a huge *Abelisaurus* sees
Ligabueino and lunges. But there is no
surprising this little meat eater.

Ligabueino sees the hunter nearing. It hears the killer grunting and its feet thumping. It smells the hunter's foul breath. With a few strides, *Ligabueino* runs away. Its speed and sharp senses have come to the rescue.

SMALL, SMALLER, SMALLEST

Compsognathus is hunting by a lake. This dinosaur looks bigger than other small dinosaurs. But this meat eater is only about the size of a turkey.

The *Compsognathus* found at the Eichstatt quarry was very small. It was probably a youngster. New discoveries show that there were dinosaurs even smaller than *Compsognathus*.

In the American West, horned dinosaurs grew to record size. Giants like *Triceratops* stretched as long as a garbage truck. Many years earlier in Asia there lived tiny horned dinosaurs.

The horned dinosaur *Liaoceratops* was only
3 feet tall. It was no bigger than a collie
dog. Even its horns were tiny. It had no nose
horn at all, just little horns under its eyes.

So which is the smallest dinosaur of all?
The smallest we have found is *Microraptor*.
Its name means "tiny thief." It was only
15 inches long from head to tail. At that
size, it could have taken only tiny animals
and insects for food.

Microraptor had feathers like a bird does. Could *Microraptor* have flown? Probably not as well as a bird. But it could flap its arms like wings. Perhaps it could hop into the air and glide.

WHERE ARE THEY?

Dinosaurs died out 65 million years ago.
Other large animals went **extinct** then too.
Many think an **asteroid** smashed into Earth.
Smoke and dust from the crash would have
blocked sunlight. These changes would
have killed dinosaurs.

Warm-blooded animals, called mammals, birds, and many small creatures survived. Perhaps small animals were able to protect themselves or had places to hide. But then why didn't tiny dinosaurs survive? No one knows.

But you could say that tiny dinosaurs are
still alive! Many scientists think just that.
They say that birds are the closest relatives
to little dinosaurs. The skeletons of birds
and small dinosaurs are very much alike.

Birds and little meat-eating dinosaurs have hollow bones. Both have many holes in their skulls. Both have curved necks, **wishbones,** and feathers. To scientists, birds are a kind of tiny dinosaur! So maybe dinosaurs are not extinct. They are flying over your head every day!

GLOSSARY

asteroid (AS-tur-oyd): a large, rocky lump that moves in space

extinct (eks-TINKT): when no members of a kind of animal or plant are living

fossils (FAH-suhlz): the remains, tracks, or traces of something that lived long ago

limestone (LEYEM-stohn): rock made of pressed lime. Lime comes from the skeletons and shells of sea creatures.

wishbones (WIHS-bohnz): collarbones. In birds, the wishbone is called the furcula.

INDEX